Louiza Oukse

Catalysts in organic chemistry

Louiza Ouksel

Catalysts in organic chemistry

ScienciaScripts

Imprint

Cover image: www.ingimage.com

This book is a translation from the original published under ISBN 978-620-6-72150-5.

Publisher:
Sciencia Scripts
is a trademark of
Dodo Books Indian Ocean Ltd. and OmniScriptum S.R.L publishing group

120 High Road, East Finchley, London, N2 9ED, United Kingdom
Str. Armeneasca 28/1, office 1, Chisinau MD-2012, Republic of Moldova, Europe
Managing Directors: Ieva Konstantinova, Victoria Ursu
info@omniscriptum.com

Printed at: see last page
ISBN: 978-613-9-55455-3

Contents

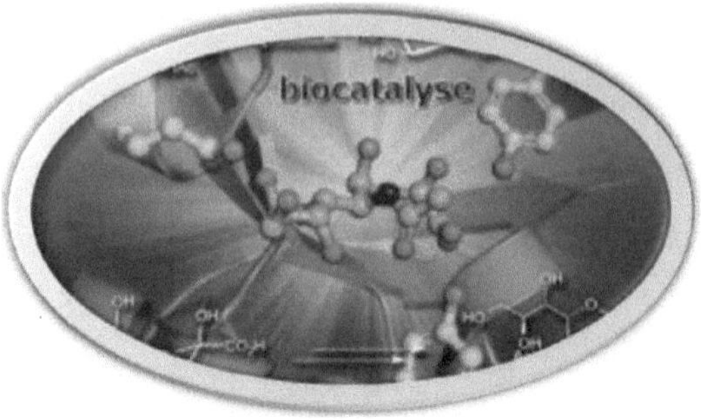
biocatalyse

Dedications

^ *I* dedicate this work to the memory of my mother, my father and my grandparents, may Allah the Almighty grant them peace and mercy.

- ✓ *To* my husband who helped me a lot with his advice.
- ✓ *To* my daughters Imen, Amina and Maissa Nour el Houda.
- ✓ *To* my brother Mohammed and my sister Zahia.

FOREWORD

The phenomenon of catalysis was first described by Berzelius in 1835. Catalysis is the increase in the speed of a reaction due to the action of a chemical species, called a catalyst, which is not involved in the reaction balance. The action of a catalyst is purely kinetic. All a catalyst does is increase, sometimes very significantly, the speed of a thermodynamically possible reaction. Under no circumstances can a catalyst make possible a reaction that is thermodynamically very unfavourable.

The catalyst accelerates the reaction by providing another pathway or mechanism for moving from the reactants to the products. This new pathway is characterised by a lower activation energy than the original pathway.

Chapter I: In this chapter we discuss homogeneous catalysis.

Chapter II: this chapter deals with heterogeneous catalysis.

Chapter III: deals with enzymatic catalysis, which is the intermediary between homogeneous catalysis and heterogeneous catalysis.

Chapter I

Homogeneous catalysis

I.1Definition

A catalyst is a species which increases the rate of a transformation, without appearing in the reaction equation and without modifying the composition of the system in the final state. In homogeneous catalysis, the reactants and the catalyst are in the same physical phase, often liquid. The kinetic study is therefore carried out using the concepts introduced in the course on reaction mechanisms. In certain simple cases, the action of catalyst C on a reaction **A ⟶ B** can be modelled as follows:

$$A + C \underset{K_{-1}}{\overset{K_1}{\rightleftharpoons}} I \qquad (1)$$

$$I \xrightarrow[K_2]{} B + C \qquad (2)$$

Reaction intermediate **I** is sufficiently reactive for the quasi-stationary state approximation to be applied. The rate of formation of product **B** is proportional to the concentration of catalyst. This explains the acceleration of the reaction in its presence. In other cases, the mechanism involving the catalyst is more complex.

$$\frac{d[I]}{dt} = V_1 - V_2 - V_3 = 0$$

$$K_1[A][C] - k_{-1}[I] - K_2[I] = 0$$

$$[I] = \frac{K1\ [A][C]}{K_{-1} + k_2}$$

The rate of formation of product B is then:

$$\frac{d[B]}{dt} = v2 = K_2[I] = \frac{K_1K_2}{K_{-1}+K_2}[A][C]$$

I.2 Acid-base catalysis

Acids and bases are the simplest of the liquid phase catalysts and the oldest known.

I.2.1 Acid catalysis

The key step is the transfer of an H^+ from the HA catalyst to the X substrate.

$$X + HA \longrightarrow HX^+ + A^-$$

$$HX^+ \longrightarrow \text{Produits}$$

Products

I.2.2 Basic catalysis

The key step is the transfer of an H^+ from the substrate XH to a base B.

$$XH + B \longrightarrow X^- + BH^+$$

$$X^- \longrightarrow \text{Produits}$$

Products

Acid-base catalysis is common in chemistry and biochemistry. The addition of a hydrogen atom or the departure of a hydrogen atom often leads to an increase in the reactivity of certain chemical functions. This is the case with the OH^-hydroxyl group. The dehydration reaction of an alcohol to an alkene takes place according to the following equation:

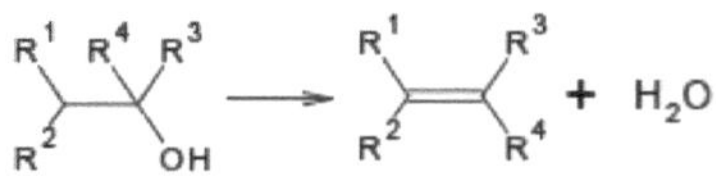

In practice, this reaction requires the C-O bond to be broken, in other words the HO group to leave in the form of an HO^-ion, since the bond is polarised -δ on the oxygen. However, the departure of a hydroxide ion is a kinetically unfavourable process, so the rate of reaction is slow. On the other hand, in the presence of an H ion$^+$, the hydroxyl group is in equilibrium with its protonated form, which allows a molecule of

molecule.

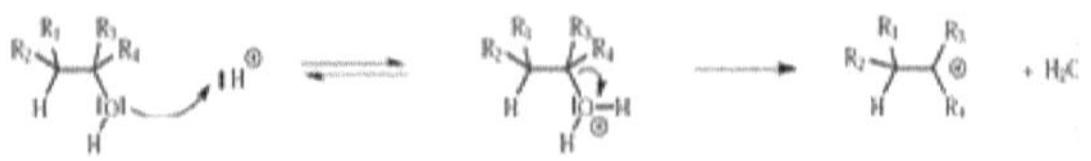

The alkene is then formed by the departure of an H^+ ion, which is favoured by the fact that the carbocation is not very stable. An H^+ ion is therefore released at the end of the reaction, which is expected for a catalyst.

I.2.3 A simple example of acid catalysis

Hydrolysis of an ester.

In a basic medium :

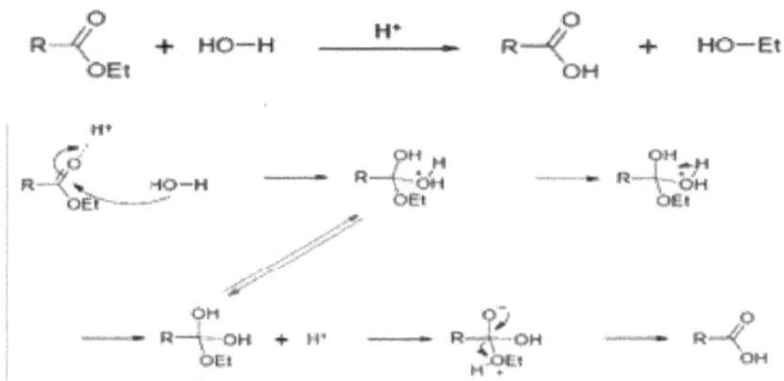

In an acidic medium, we have the mechanism:

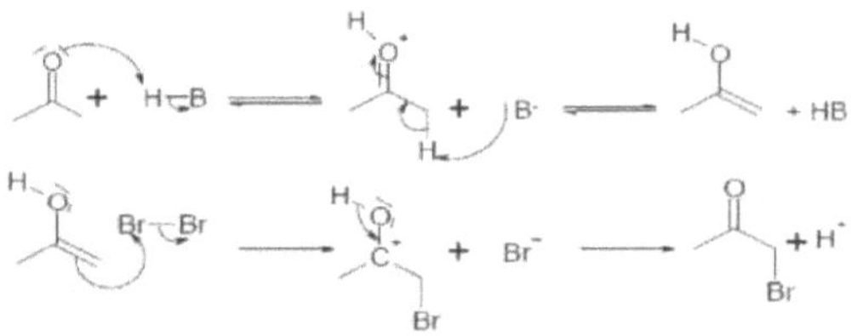

In a basic medium, we have the mechanism:

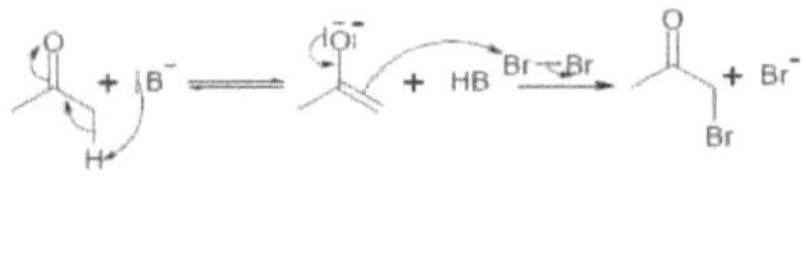

This is a specific acid-base catalysis, meaning that the speed of the reaction depends only on the pH of the solution. For an acid-catalysed reaction, there are two possible cases:

> Specific acid catalysis

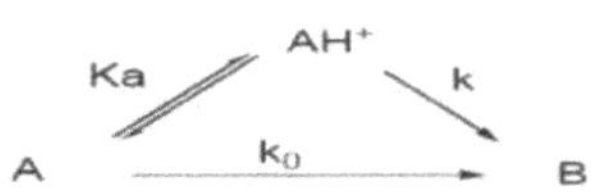

Equilibrium is reached between AH^+ and A.

The rate law is written :

$$v = k_0[A] + k[AH^+] = k_0[A] + kK[A][H_3O^+]$$
$$v = (k_0 + kK[H_3O^+])x[A]$$

Hence

$$k = k_0 + k_{H_3O^+}[H_3O^+]$$

> Generalised acid-base catalysis

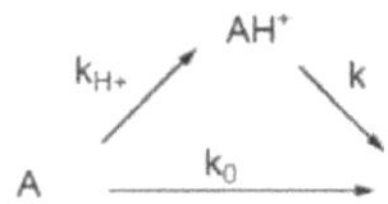

Equilibrium is not reached between AH^+ and A and k_{H^+} depends on the acid in solution. So, for a mixture of acids and bases, we obtain a general law :

$$k = k_0 + k_{H_3O^+}[H_3O^+] + k_{HO^-}[HO^-] + \sum_i \left(k_{HA_i}[HA_i] + k_{A_i^-}[A_i^-]\right)$$

where HA_i and A_i^- are the different acids and bases in solution.

1.3 Auto-catalysis

A reaction is autocatalysed when one of the reaction products is involved in the rate equation. An example of such a reaction is the oxidation of oxalic acid by potassium permanganate.

▶ $\mathbf{2MnO_4^- + 5C_2O_4^{2-} + 16H^+ \rightarrow 2Mn^{2+} + 10CO_2 + 8H_2O}$

The decolourisation of the permanganate is slow at the start of the reaction and then becomes faster. The kinetic study shows that the speed of the reaction is of the form :

$$\vartheta = k\,[C_2O_4^{2-}][Mn^{2+}]$$

The Mn^{2+} ions autocatalyse the reaction.

Another example is the hydrolysis of esters catalysed by the weak acid released during the reaction.

▶◀ $\mathbf{HCOOC_2H_5 + H_2O \longleftrightarrow HCOOH + C_2H_5OH}$

At the start of the reaction, the reverse reaction is not taken into account.

$$\vartheta = k\,[HCOOC_2H_5][HCOOH]$$

1.4 Homogeneous catalysis by transition metal complexes

1.4.1 Transition metals

A transition metal is a chemical species that forms one or more stable ions with partially filled **d** orbitals. They are widely used for their reactivity in catalysis and as active centres in biological chemistry (metalloproteins). The main reasons why transition metals contribute to catalytic systems can be summarised as follows:

- ✓ Binding capacity.
- ✓ Choice of ligands.
- ✓ Ligand effects.
- ✓ Oxidation state variability.
- ✓ Their coordination numbers.

1.4.2 Coordination complexes

In an empirical representation, **M.L.H. Green** proposes to represent any coordination complex in the general form.

$$\mathbf{ML_nX_xZ_z}$$

With

M: transition metal, surrounded by ligands L, X, Z, **L:** ligand donating an electron pair to the bond (Lewis base) CO, $R_1R_2C{=}CR_3R_4$, NH_3, NR_3, PR_3, $P(OR)_3$, OH_2, OR_3, SR_2

X: ligand that donates an electron to the bond, H, CR_3(R=H, alkyl, aryl,vinyl),Ar,C(O)R, SiR_3, NO, NR_2, PR_2,OH,RCOO,OOH, NO_2, F,Cl,Br,I,CN.

Z: ligand that is not an e^-donor (and therefore accepts two electrons) in the bond (Lewis acid) H^+, BX_3, AlX_3

A complex is a polyatomic structure consisting of a central atom or cation to which molecules or ions called ligands are bound. It is characterised by its geometry, which depends on its coordination index. The formation of a complex involves a reaction between the ligands (which contribute electron doublets) and the metal or metal ion (which contributes electron doublets).

metal or metal ion (which has empty orbitals)

which leads to the formation of a covalent bond of

bond. The type of hybrid orbital formed depends on the geometry of the complex.

Number of coordination	Geometry	Hybrid orbital	Example
2	Linear	sp or ds	$[Cu(NH_3)_2]^+$
4	Tetrahedral	sp^3or d^3s	$[Zn(NH_3)_4]^{2+}$
4	Flat square	dsp^2	$[Ni(CN)4]^{2-}$
6	Octahedral	d^2sp^3or $sp^{(3)\,d2}$	$[Cr(NH_3)_6]^{3+}$

Example:

$[Cu(NH_3)_2]^+$ Cu^+:$[Ar]3d^{10}4s^04p^04d^0$

$[Zn(NH_3)_4]^{2+}$ Zn^{2+}:$[Ar]3d^{10}4s^04p^04d^0$

$[Ni(CN)_4]^{2-}$ Ni^{2+}:$[Ar]3d^84s^04p^04d^0$

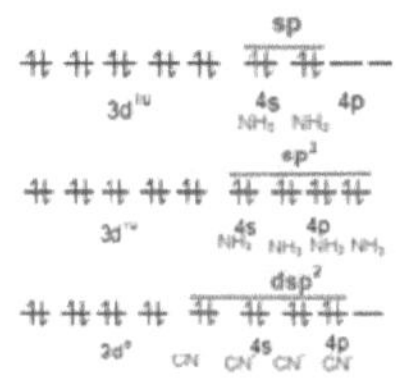

1.4.3 Elementary stages and the catalytic cycle

A catalytic reaction can be broken down into a number of elementary steps, for example :

-*S* Activation by coordination of the substrate and the reagent, either simultaneously or successively.

S Reaction between activated species.

S Decoordination of the product(s).

Insofar as the metal complex is in the same state at the end of the third stage as it was at the start of the first, the process can, in principle, be repeated indefinitely, the complex describing a sequence of transformations which constitutes the catalytic cycle, each cycle corresponding to a rotation of the catalyst. If, at the end of the last stage, the metal complex is in a different state to that at the start, auxiliary regeneration reactions can still be used to return it to its initial state.

1.4.4 The role of ligands in catalysis

The coordination of the reagent and substrate around the metal in the active complex presupposes the existence of vacant sites. These can be created by the dissociation of one or more ligands, which must therefore have suitable properties. As well as protecting potential coordination sites, the additional ligands help to maintain the complex in a soluble form, stabilising the metal at an often low valency and preventing it from changing to an insoluble metallic state. An example of this is in the hydroformylation of olefins, where the **$HCo(CO)_4$** catalyst is stabilised against thermal decomposition by carbon monoxide. Additional ligands also influence the reactivity of the complex and the selectivity of the catalytic reaction through their electron donor or acceptor properties, their steric hindrance, their chelating power, etc.

a) Implementation

By definition, coordination catalysts are used in a homogeneous liquid phase. The reactors used can operate in a static regime (batch reaction) or in a dynamic regime (continuous reaction), with the reactor full or partially filled (operation in liquid-vapour equilibrium). In a dynamic system, the catalyst is generally also injected continuously in liquid form, either pure or dissolved in a solvent which may be the reactant or the reaction product. It is usually withdrawn continuously with the reaction effluent, but if the effluent is collected in the vapour phase by boiling the reaction mixture, the catalyst can accumulate in the reactor, from which it is withdrawn at variable time intervals. When the catalyst becomes soluble in the reaction effluent, it generally has to be separated from it, to obtain a product that is not polluted by sometimes toxic metallic residues. If the catalyst is inexpensive and used in small quantities, it can be discarded (known as a "lost catalyst" operation).

b) Estimating the performance of a catalyst

From both an applied and a fundamental point of view, it is necessary to have criteria for assessing the performance of a catalyst in a given reaction. First of all, we use the usual criteria for any chemical reaction:

S Conversion rate of the reactant(s).

S Selectivity of the desired product and yield.

In any case, the best way of expressing the activity of a catalyst remains the rate constant of the catalytic reaction.

c) Reactions

S Isomerisation

We will consider here essentially the positional isomerisation of the double bond of olefins (alkenes), which is most often accompanied by *cis-trans* rebalancing.

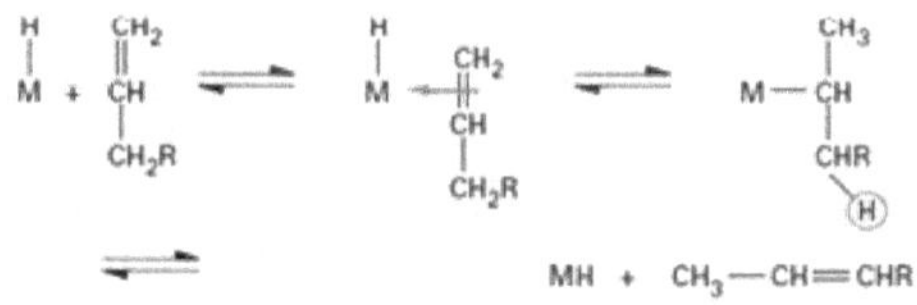

✓ *Hydrogenation*

The hydrogenation of alkenes by coordination catalysis is one of the reactions that has been most extensively studied and whose mechanisms are best understood.

$$MnX + H_2 \rightarrow MnH + HX$$

- Hydrogenation of alkenes.
- Hydrogenation of monoalkenes.
- Hydrogenation of functionalised alkenes.
- Hydrogenation of acetylenic compounds.
- Hydrogenation of conjugated dienes.
- Asymmetric hydrogenation of ketones and nitrile functions.

✓ *Dimerisation, oligomerisation and polymerisation*

The condensation of an unsaturated hydrocarbon onto itself occurs in the presence of various catalytic systems and, depending on the degree of condensation, is called :

- **Dimerisation**, when condensation stops at the second monomer unit.
- **Oligomerisation**, when the number of condensed monomer units is low, but greater than

two.

- **Polymerisation**, when the number of condensed

monomer units is high.

Oligomers

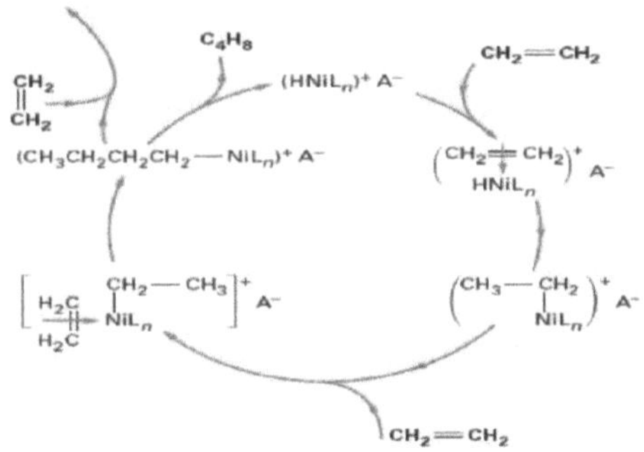

Example 1: Dimerisation of ethylene

Degenerate polymerisation mechanism. Many complexes of palladium, rhodium and especially nickel respond to this type of mechanism. Dimerisation reactivity decreases with this type of catalyst as the degree of substitution of the double bond increases:

Ethylene>> propene> butenes> isobutene

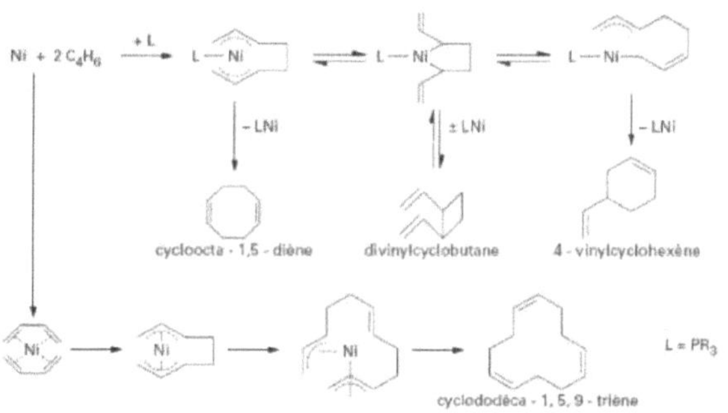

Example 2: Cyclodimerisation and trimerisation of butadiene

Two groups of catalysts are capable of carrying out this type of transformation.

- Ziegler systems based on titanium and zirconium give a range of oligomers with a generally broad distribution (up to C30, with a maximum around C10-C12).
- Better results are obtained with nickel-based catalysts co-ordinated by chelating phosphine ligands, such as :

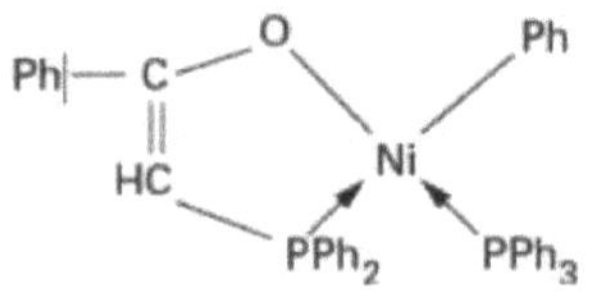

→ *Alkene polymerisation*

Alkene polymerisation catalysts are generally compounds of metals from groups 4 to 6 of the periodic table, especially titanium, zirconium, vanadium and chromium. Homogeneous polymerisation catalysts are generally only active on ethylene or ethylene-propene mixtures.

Example 3: Polymerisation of dialkenes

This is in fact limited to the polymerisation of 1,3-butadiene and, to a lesser extent, isoprene. Three types of structure can be obtained: *cis-1*,4, *trans-1*,4 or -1,2 in the case of polybutadiene.

The metathesis (or disproportionation) reaction of an unsaturated substrate results in the simultaneous cleavage of the σ and π bonds of the double (or triple) bond, with redistribution of the alkylidene (or alkylidyne) radicals. The reaction between linear alkenes leads to new linear alkenes. Metathesis of a cyclic alkene leads to ring opening at the unsaturation point, with the formation of an unsaturated polymeric chain. In addition, the cross-reaction between a linear alkene and a cyclic alkene gives unsaturated telomers.

> Disproportionation reaction of acyclic alkenes.

$$\underset{50\%}{2\,R_1CH{=}CHR_2} \rightleftharpoons \underset{25\%}{R_1CH{=}CHR_1} + \underset{25\%}{R_2CH{=}CHR_2}$$

> Polymerisation of cyclic alkenes.

> Telomerisation of cyclic alkenes by acyclic alkenes.

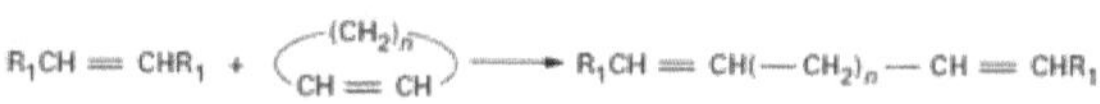

Example 4: Metathesis reactions

Metallacyclobutane mechanism.

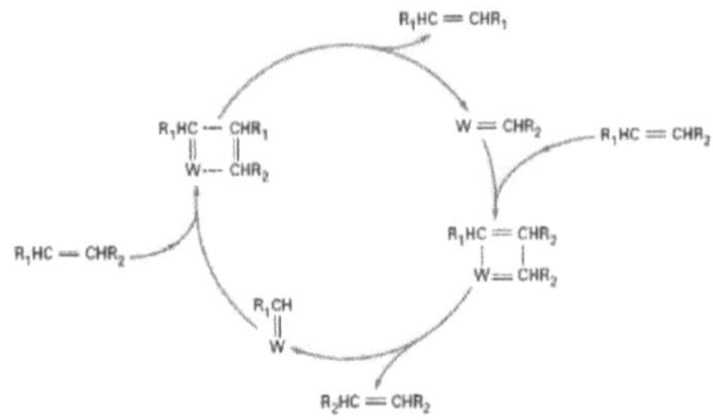

Example 5: Carbon-carbon coupling reactions

- *Heck reaction*

The coupling reaction between a halogenated vinyl or aryl compound and an alkene leads to the formation of a 1,3-diene or a substituted styrene, according to :

$$Ar\text{-}X + R^1\text{—}CH{=}CH_2 \longrightarrow R^1\text{—}CH{=}CH\text{-}Ar + HX$$

$$R\text{—}CH{=}CH\text{-}X + R^1\text{—}CH{=}CH_2 \longrightarrow R\text{—}CH{=}CH\text{—}CH{=}CH\text{—}R^1 + HX$$

The reaction, discovered by Heck and Mizoroki, is catalysed by palladium complexes. The palladium complexes used are very varied, but often include a phosphine ligand. They can be generated in situ from a palladium salt and the ligand.

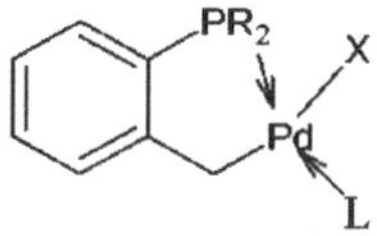

- *Suzuki, Stille and similar reactions*

Here we look at coupling reactions between a halogenated compound (or a similar electrophile, such as a triflate) and organometallic compounds. These reactions are represented by the following general equation :

$$R\text{—}M + R^1\text{—}X \longrightarrow R\text{—}R^1 + M\text{—}X$$

These reactions are catalysed by nickel complexes and, above all, palladium complexes, the latter giving better selectivity.

They correspond to the general mechanism :

$$Pd(0) + R^1\text{—}X \longrightarrow R^1\text{—}Pd(II)\text{—}X$$

$$R^1\text{—}Pd(II)\text{—}X + R\text{—}M \longrightarrow R^1\text{—}Pd(II)\text{—}R + MX$$

$$R^1\text{—}Pd(II)\text{—}R \longrightarrow R^1\text{—}R + Pd(0)$$

Chapter II

Heterogeneous catalysis

11.1 Introduction

The kinetics of chemical reactions occurring in heterogeneous chemical systems have specific characteristics. The aim of this book is to specify these specificities, to develop their consequences and to propose a methodology for analysing the mechanisms and kinetics of heterogeneous reactions.

The heterogeneous nature of the systems under consideration means that many of the processes and reaction stages take place at the interfaces that delimit and separate the various phases of the reaction system. The role and influence of interfacial steps and processes are fundamental elements of heterogeneous kinetics. This influence is also reflected in the very important role played by the shape and size of the reactants and reaction products, which determine the extent of the interfacial zones and any variations in them. Thus, the kinetic analysis of reactions and transformations occurring in a heterogeneous system will, in general, have to take into account a set of specific processes such as:

1-Transfer of reagents to the solid surface.

2-Adsorption of reactants onto the catalyst.

3-Interaction between reactants adsorbed on the surface. 4-Desorption of products from the catalytic surface. 5-Transfer of products leaving the catalyst.

Stages 2, 3 and 4 are chemical in nature, whereas stages 1 and 5 correspond to physical matter transfer processes.

11.2 Diffusion of reactants and products

Diffusion, a physical process, tends to equalise concentrations in the fluid phase. A diffusion current is therefore established in the vicinity of the solid-fluid interface, when the consumption of reactants on the surface and the formation of products cause concentration gradients in this region.

11.2.1 External diffusion

The outside of the grain represents the entire surface of the catalyst (plus the intergranular surface).

11.2.2 Internal diffusion

The outside of the grain represents only a small fraction of the surface of the catalyst if it is porous.

The internal surface of the catalyst is made up of channels, access to which is always controlled by diffusion.

11.3 Adsorption of reagents onto the catalyst

Once they reach the immediate vicinity of the catalytic surface, the reagents will adsorb to it, i.e. react with it to give rise to new, more reactive chemical species. Adsorption therefore often results in the breaking of intramolecular bonds, and bonds are always formed with the surface.

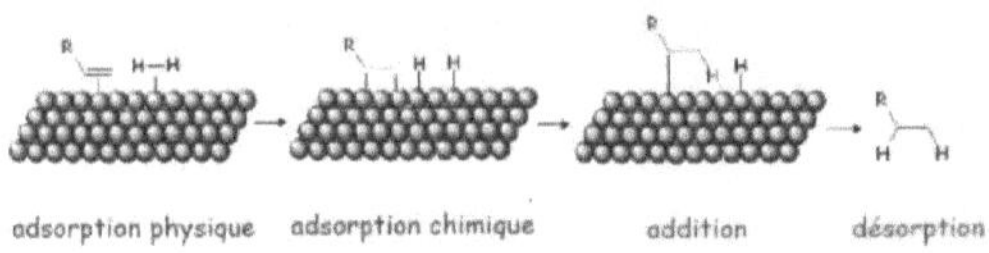

II.1.2.1 Chemical adsorption

Chemisorption of molecules M on surface sites **S** gives rise to an adsorption equilibrium :

$$M + S \rightleftharpoons M\text{-}S$$

This can be studied thermodynamically at temperatures below that of the catalytic reaction. The state of the adsorbed molecules varies greatly, depending on whether they are dissociated or not, but in all cases the number n of adsorbed molecules per unit mass of catalyst increases with the pressure P of **M** (or with its concentration C in the liquid) to reach a limit value n OT corresponding to saturation of all the **S** sites. We then define the surface coverage rate :

$$\theta = \frac{n}{n_\infty} \quad (\theta \leq 1)$$

Adsorption-desorption isotherms express the variation of θ as a function of P (or C) at constant temperature, for an adsorbate-adsorbent system. Adsorption energies are measured directly by microcalorimetry, or determined indirectly, using a network of isotherms measured over an entire temperature range. The isosteric heat of adsorption qst for one mole of M is most frequently used. From the measurement of the equilibrium pressure *P* (*T*) for a constant adsorbed quantity *n*, we can deduce :

$$\left[\frac{d(lnP)}{dT}\right]_n = \frac{q_{st}}{RT^2}$$

The heat of adsorption is a positive quantity, of the order of 50 to 400 kJ per mole. Thus, for a bare nickel surface, qst is 125 kJ for H_2, 145 kJ for CO and 540 kJ for O2 .

$$qst = -\Delta Ha$$

Langmuir's model is the simplest. It applies to a homogeneous surface, i.e. one for which the heat of adsorption is independent of the coverage rate θ.

S: surface covered by an adsorbed monomolecular layer

50: free surface

ST: total surface area

ST: S + S0

Adsorption Desorption

$$V_{ads}= k_{ads}P\ S_0 \qquad V_{des}= k_{des}S =k_{des}(S_T - S_0)$$

At dynamic equilibrium :

$V_{ads} = V_{des}$

$k_{ads}P\ S_0 = k_{des}(S_T - S_0)$ (1)

The fraction of surface covered is called Θ.

$$\theta = \frac{S}{S_T} = \frac{S_T - S_0}{S_T}$$

Hence

$$S_T - S_0 = \theta S_T$$

The fraction of free surface is then defined as 1- Θ

$$1 - \theta = \frac{S_0}{S_T}$$

hence

$$S_0 = (1 - \theta)S_T$$

By replacing in equation (1), we have :

$$k_{ads}\ P\ (1-0)\ S_T = k_{des}\ 0\ S_T$$

Experience shows that adsorption in certain cases is accompanied by a bond break or the opening of a multiple bond at the level of the gas molecule. The atoms then attach themselves to neighbouring sites. Adsorption is said to be dissociative.

On the other hand, when several gases present can be simultaneously chemisorbed on the same type of site, competition is established. Each gas has its own rate of adsorption and desorption. The sites occupied by a molecule of gas A are no longer accessible to a molecule of gas B. This is known as competitive adsorption.

Heterogeneous catalysis can only involve chemisorbed molecules. The chemical reaction only affects the reagent attached to the surface of the catalyst. The speed of the reaction is therefore proportional to the number of reactant molecules attached to the surface of the catalyst, i.e. σ - G Let vm be the speed for a mass m of catalyst with q the speed constant at the surface of the catalyst. ***V is*** then defined as the rate of catalytic reaction per unit mass of catalyst. With **k** the rate constant of the catalytic reaction referenced to the unit mass of the catalyst.

In the case of competitive adsorption, we have seen :

If the gases are weakly chemisorbed λ_A and λ_B are small

$\lambda_A P_A$ **et** $\lambda_B P_B <<< 1$

$V = k\ \lambda_A P_A\ \lambda_B P_B\ = k'\ P_A\ P_B$

Velocity order 1 / A and order 1 / B Overall order = 2

If the gas (A) is much more adsorbed than the other (B) Velocity order -1 / A and order 1 / B

Reagent A, by its massive occupation of the catalyst surface, prevents the reaction between the 2 chemisorbed reagents.

II.2 Hydrogenation catalysts

II.2.1 Definition

Hydrogenation is a chemical reaction involving the addition of a molecule of dihydrogen (H_2). This reaction is usually used to reduce or saturate organic compounds. This reaction generally requires catalysis, as reactions without catalysis require very high temperatures. The hydrogenation reaction is useful in many fields, particularly in petrochemistry (transformation of alkenes into alkanes) and in the food industry (hydrogenation of unsaturated fats into saturated fats).

a) Re, Ru, Co, Ni, Pt, Rh

Although the reaction is thermodynamically favourable, it is difficult to start due to the high stability of dihydrogen. Thus, apart from a few exceptions, no reaction occurs between H_2 hydrogen and the organic compound at a temperature below 480°C and in the absence of a catalyst. On the one hand, heating the reaction medium to high temperature is impractical (although feasible), and on the other, some organic compounds cannot withstand high temperature. The preferred solution is therefore to use a catalyst. This catalyst will bind both the dihydrogen and the unsaturated compound, and so 'help' them to react together. Metals from the platinum group, in particular platinumplatinum oxide, , palladium, palladium and can be used as catalysts.

Catalysts in organic chemistry Rhodium and ruthenium are particularly active catalysts, operating at low temperatures and low H_2 pressures. Ruthenium is a versatile element in both heterogeneous and homogeneous catalysis, but it is in the latter field that it stands out for the applications it has developed in catalysis.

The Wilkinson reagent catalyses the hydrogenation reaction of alkenes. Unlike other catalysts commonly used for this type of reaction (Raney nickel, Lindlar catalyst, etc.), it is a homogeneous catalyst. The mechanism begins with the dissociation of one or two triphenylphosphine ligands to provide 12 or 14 complexing electrons, followed by an oxidising addition of dihydrogen to the metal. This is followed by complexation π of the alkene, intramolecular transfer of hydrogen to the alkene and then reductive elimination to liberate the alkane formed.

Certain soluble platinum, palladium or rhodium-based catalysts can be used to react catalytic hydrogenation in a homogeneous phase. This is the case with the Wilkinson-Osborn catalyst: $RhCl(PPh_3)_3$

The reaction proceeds by a mechanism involving oxidative addition and reductive elimination, with a hydrometallation step that leads to stereoselectivity.

- Oils can be transformed into fats by catalytic hydrogenation (on the surface of metals such as Ni, Pt or Pd). This operation is used to obtain margarines from vegetable oils.
- Asymmetric hydrogenation: Hydrogenation can be carried out in the presence of chiral ligands, which transmit their chirality to the products formed. Catalysts are based on rhodium, ruthenium, platinum and palladium.

b) Copper oxide, chromium oxide

These Cu^{+} and Cr^{3+}ions are electron acceptors that give the non-strechiometric Cu_2O and CPO.í oxide P-type semiconducting properties.

c) Mixed noble metal catalysts

Formation of cyclic and aromatic molecules from alkanes on platinum or platinum alloys (Pt-

Re, Pt-Sn).

II.2.2 Applications

Hydrogenation has many industrial applications. These include the Haber process for hydrogenating nitrogen to ammonia , and the Bergius process for producing synthetic petrol from coal. Hydrogenation is also used in the petrochemical industry to convert alkenes into less volatile alkanes, and aromatic compounds into saturated alkanes (paraffins) and cycloalkanes (naphthenes).

a) Fischer-Tropsch

The Fischer-Tropsch process is a reaction involving carbon monoxide and dihydrogen to convert them into hydrocarbons. The catalysts used for this reaction are usually based on iron or cobalt. The aim of conversion is to produce synthetic liquid fuel from coal and gas. *b) Hydroformylation / hydrocarboxylation*

This involves the addition of a hydrogen (hydro-) and a CHO (formyl) group to a double bond, catalysed by rhodium or cobalt complexes.

c) Synthesis of menthol (Tagasako company)

The process begins by forming an allylic amine from myrcene, which undergoes asymmetric isomerisation in the presence of a BINAP rhodium complex to give (after hydrolysis) enantiomerically pure R-citronellal. The result is treated with zinc bromide to give isopulegol which is then hydrogenated to give pure (1R, 2S, 5R)-menthol.

II.3 Catalyst inhibitors and poisons

Catalyst poison is any substance that reacts with an active site on the surface of a catalyst, partially or totally blocking its action. Such substances can be responsible for the failure of catalytic hydrogenation, but can also be used to make the catalyst less reactive and more selective. This is the case, for example, with the Lindlar catalyst, a heterogeneous catalyst in which certain sites are deactivated by the action of lead acetate . The "classic" poisons are sulphur compounds (thiols, for example, whose negative effect on the catalyst increases with the length of the carbon chain). Basically, catalyst poisons can be classified into three groups:

- compounds containing non-metallic elements such as nitrogen, phosphorus, arsenic, antimony, sulphur, selenium, tellurium, bromine, iodine and sometimes chlorine.
- Metal compounds and ions. For example Cu^{2+}, Cu^{+} , Ag^{+} , Au^{+} , Zn^{2+}, Cd^{2+}, $Hg2^{2+}$, $Hg^{(2+})$, Mn^{2+}, $Fe^{(2+})$, $Co^{2+}$and $Ni^{2+}$are poisons for platinum catalysts.
- complexing compounds metals, for example cyanide ions, carbon monoxide or benzene.

a) Hydrogenation catalyst promoters

A promoter is a chemical species which, on its own, has no effect on the reaction, but which reinforces the action of the catalyst. For example, the synthesis of ammonia NH3 from N2 and H2 is catalysed by metallic iron. Alumina Al2O3 is a promoter of the reaction: its presence disperses the iron particles, increasing the surface area of the available catalyst. *b) Hydrogenation processes*

Hydrogenation is a chemical reaction in which a hydrogen atom is added to a hydrocarbon molecule in the presence of a catalyst. Hydrogenation may or may not be destructive. *c) Hydrogenation equipment and conditions*

The use of hydrogen requires certain precautions to avoid obtaining explosive hydrogen/air mixtures. Typically, the hydrogenation reactor must be pressure tested and then inerted by nitrogen sweeps before being introduced. Similarly, at the end of the reaction, the installation is purged with nitrogen to ensure that it remains in stable conditions. Hydrogenation requires a constant pressure to be maintained. The principle of hydrogenation tends to be highly exothermic, implying the need to control the temperature. The equipment generally used for hydrogenation processes is : The T800 panel supervisor and 21 CFR Part 11.

d) Hydrogenation of some organic functions (alkenes, alkynes, nitrogen compounds)

→ **Alcènes**

$$R_2C{=}R_2C \xrightarrow[\text{Pt,Pd,Rh}]{H_2} R_2CH{-}R_2CH$$

→ **Alcynes**

$$RC{\equiv}RC \xrightarrow[\text{Pt/Pd}]{H_2} RHC{=}CHR$$

→ **Nitriles**

$$RCN \xrightarrow[\text{Ni,Rh}]{H_2} RCH_2NH_2$$

II.5 Hydrogenolysis catalysts

Hydrogenolysis is a chemical reaction by which a covalent bond carbon-carbon or carbon-heteroatom is decomposed or undergoes lysis by the action of hydrogen . The heteroatom is most often oxygen , sulphur or nitrogen . A related reaction is hydrogenation which involves the addition of a molecule of dihydrogen (H_2) without breaking a covalent bond. This reaction generally requires catalysis , with non-catalysed reactions requiring very high temperatures, and is generally carried out using hydrogen gas.

$$R\text{-}X + H_2 \rightleftharpoons R\text{-}H + H\text{-}X$$

$$R^1\text{-}R^2 + H_2 \rightleftharpoons R^1\text{-}H + R^2\text{-}H$$

a) Hydrogenolysis (benzyl ethers, ethers, alcohols)

When benzyl ethers or alcohols are subjected to the action of hydrogen in the presence of a metal catalyst, the benzyl carbon-oxygen bond is broken, and such a transformation is an example of hydrogenolysis.

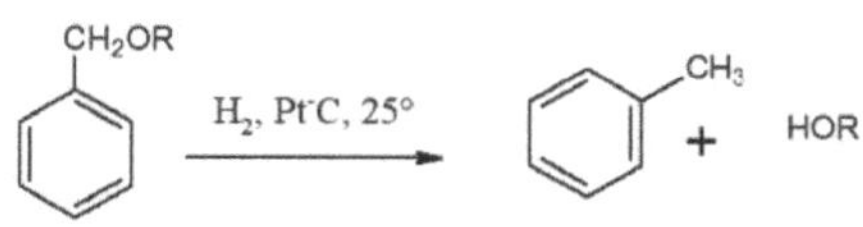

Diagram of the cleavage of benzyl ethers by hydrogenolysis. Hydrogenolysis is not possible with alcohols and ordinary ethers, so the phenylmethyl (benzyl) substituent represents a valuable protective group for hydroxyl functions; the following diagram illustrates its application in part of the synthesis of a compound (belonging to the eudesmane class) found in various essential oils and of particular importance both as a medicine and as a fragrance.

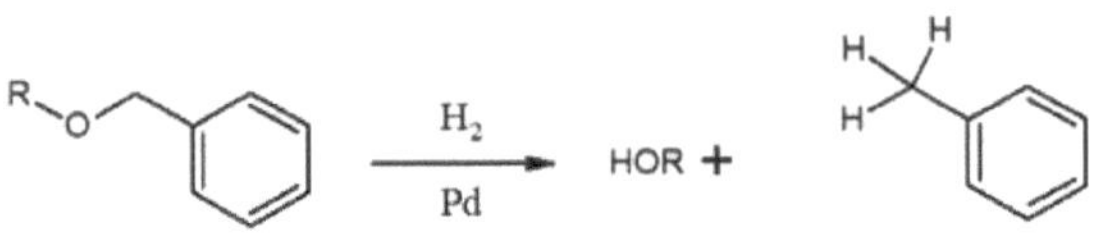

A first thioketal is produced by reacting the ketone with an appropriate thiol. The product is then transformed into an alkane by hydrogenolysis, using Raney nickel.

b) Hydrogenolysis of carbonyl compounds

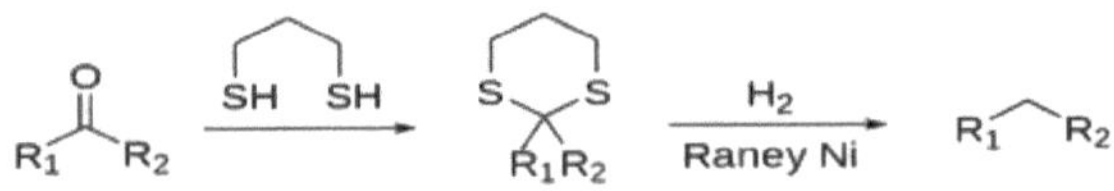

Carbonyl groups can be converted to methylene groups by desulphurising thioketals. The cyclic thioketal derived from ethanedithiol is commonly used. Reaction with excess Raney nickel leads to hydrogenolysis of the two C-S bonds.

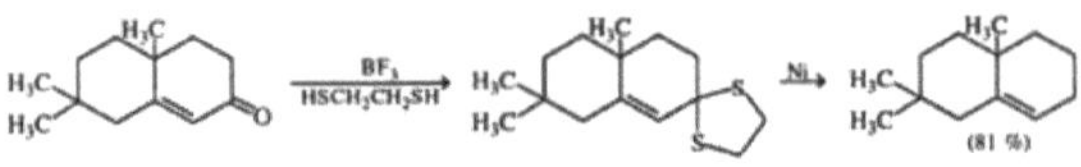

d) Hydrogenolysis of aromatic compounds

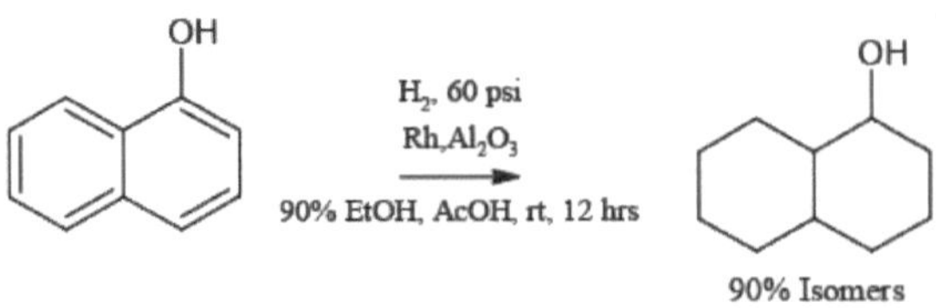

d) Hydrogenolysis of heterocyclic compounds

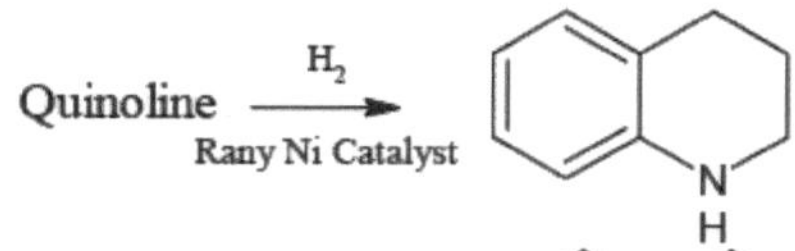

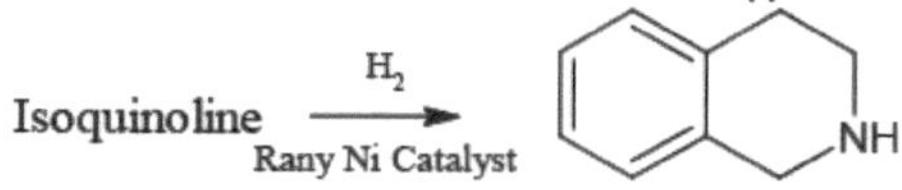

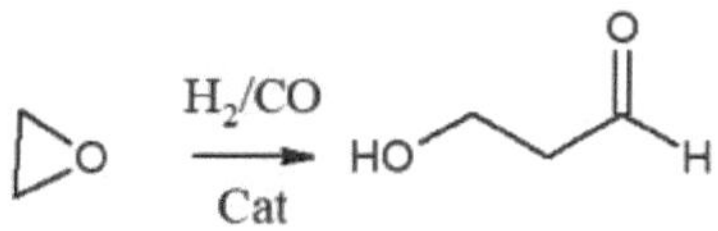

II.6 Oxidation catalysts and their applications *a) Oxidation of elements (1st, 2nd, 3rd, 4th, ...)*

> Alkanes

Alkaline potassium permanganate and chromic acid have no effect on alkanes when cold. When hot, they attack unsaturated carbons with a tertiary carbon to give first an alcohol (isobutane $(CH_3)_3CH$ gives *t-butanol* $(CH_3)_3C\text{-}OH$) and then a mixture of acids. It should also be noted that certain micro-organisms (*Methylococcus capsulatus*) possess enzymes capable of oxidising alkanes to alcohols.

> Alkenes

-Oxidation without breaking the double bond (mild)

This oxidation is *trans-regioselective* with respect to the largest substituent on the double bond. *-Oxidations with breakage of the double bond (strong)*

Oxidation with breakage of the double bond occurs with ozone and with chromic anhydride or permanganate in an acid medium.

> Alkynes

-Oxidation without breaking the triple bond (mild)

Ruthenium tetroxide, selenium oxide, SeO_2, in the presence of catalytic quantities of sulphuric acid, potassium permanganate in a neutral medium, iodine in dimethyl sulphoxide (DMSO), or thalium(III) nitrate in mild conditions, oxidise internal alkynes to α-diketones.

-Oxidation with breakage of the triple bond (strong)

Alkynes are generally more resistant to oxidation than their corresponding alkenes. Hot potassium permanganate or ozone in an aqueous medium oxidises alkynes by supplying two acids. > Alcohols

Aliphatic primary alcohols are oxidised to aldehydes and secondary alcohols to ketones by a large number of oxidants.

Chapter III

Enzymatic catalysis

III.1Definition

Enzymatic catalysis is a catalytic process specific to living beings in which enzymes, complex protein molecules with a high molecular mass, play the role of catalysts.

There are six categories of catalytic mechanisms used by enzymes:

- Acid-base catalysis.
- Covalent catalysis.
- Metal ion catalysis.
- Electrostatic catalysis.
- Proximation and orientation catalysis.
- Catalysis by preferential binding to the transition state complex.

III.1 Enzyme classification and nomenclature

111.1.1 Classification

1) Oxido-reductases: Catalyse oxidation-reduction reactions.
2) Transferases: Transfer a functional group (e.g. a methyl or phosphate group).
3) Hydrolases: Catalyses the hydrolysis of various bonds.
4) Lyases: Break various bonds by processes other than hydrolysis and oxidation.
5) Isomerases: Catalyse isomerisation reactions in a single molecule.
6) Ligases: Join two molecules by covalent bonds.

111.1.2 Nomenclature

The EC nomenclature (EC stands for Enzyme Commission numbers) is a numerical classification of enzymes, based on the chemical reaction they catalyse. Each enzyme code consists of four numbers separated by dots. Consists of the capital letters "EC" followed by four digits. Each number represents a step in the enzyme's classification.

It is written in the form: **E.C. X.X.X.X.**

- The first X corresponds to the enzyme class.
- The second X allows further classification according to the nature of the electron donor

group on which the enzyme acts.

- The third X allows further classification according to the nature of the electron acceptor group on which the enzyme acts.

- The fourth X allows further classification according to the substrate on which the enzyme acts.

The first X indicates the type of reaction catalysed, i.e. the class of enzyme.

N°	**Enzyme class**
X=1	Oxidoreductases (electron transfer: redox)
X=2	Transferases (transfer of a functional group from one molecule to another)
X=3	Hydrolases (bond cleavage by hydrolysis)
X=4	Lyases (bond cleavage by elimination)
X=5	Isomerases (change in configuration of substrate molecule)
X=6	Ligases (condensation of two molecules)

The second X indicates the nature of the substrate.

class of hydrolases (**3**): 3.1 Acting on ester bonds 3.2 Glycosylases

3.11 Acting on Carbon-Phosphorus Bonds

3.12 Acting on Sulfur-Sulfur Bonds

3.12 Acting on Carbon-Sulphur Bonds

In *hydrolases acting on ester bonds* (**3.1**) :

3.1.1 Carboxylic Ester Hydrolases

3.1.2 Thiolester Hydrolases

3.1.3 Phosphoric Monoester Hydrolases

3.1.30 Endoribonucleases Active with either Ribo- or Deoxyribonucleic Acids and Producing 5'-Phosphomonoesters

3.1.31 Endoribonucleases Active with either Ribo- or Deoxyribonucleic Acids and Producing 3'-Phosphomonoesters

In *carboxyl ester hydrolases* (**3.1.1**) :

3.1.1.1 carboxylesterase

3.1.1.2 arylesterase

3.1.1.3 triacylglycerol lipase

3.1.1.78 polyneuridine aldehyde esterase

3.1.1.79 hormone-sensitive lipase

Example: pancreatic lipase, which hydrolyses triglycerides, has the following nomenclature:

E.C.3.1.1.3

III.1.3 Efficacy of an enzyme and conditions

> **Temperature**

The enzyme is capable of catalysis at a temperature equal to that of living organisms (37°C-38°C). At higher temperatures, the enzyme will be denatured, leading to modification of the active site. The reaction cannot therefore take place. At low temperatures (below 5°C), the enzyme is inactivated because the molecular agitation caused by heat is limited: it is difficult for the substrate molecules and enzymes to meet.

> **pH**

The enzyme is not very sensitive to pH. The enzyme functions at neutral, acidic and basic pH. However, a pH that is too high (between 13 and 14) leads to denaturation of the enzyme.

III.1.4 Enzymatic kinetic law

Consider the enzymatic reaction:

$$E + S \underset{k_{-1}}{\overset{k_1}{\longrightarrow}} ES \overset{k_2}{\longrightarrow} E + P$$

E: Enzyme, S: Substrate

ES: Enzyme-substrate complex, P: Product

The enzymatic reaction obeys the **Michaelis-Menten** law.

$V_i = (V_{max} * [S]) / (K_m + [S])$

Vi: Initial speed (i.e. in the absence of product) of the enzymatic reaction for a substrate concentration [S], Vmax: Maximum initial speed measured for a saturating substrate concentration [S].

[S]: Substrate concentration.

Km: Specific Michaelis constant for the enzyme. This is the substrate concentration for which the initial reaction rate is equal to Vmax/2.

III.2 Uses of enzymes

111.2.1 Industrial uses.

Since enzymes are soluble in water and very fragile, one industrial process consists of immobilising them on a solid support by adsorption, inclusion, cross-linking or chemical fixation. In this way, they can catalyse reactions in a heterogeneous environment that could be catalysed in a homogeneous phase.

Examples of reactions

- Saccharification of starches by amylases.
- Production of glucose from cellulose by cellulase.
- Hydrolysis of lactose by lactase.
- Hydrolysis of fats.

111.2.2 Household uses.

Detergent powders contain mixtures of proteases which work effectively at low temperatures to rewash the detergent.

111.2.3 Other uses.

Digestive enzymes, which, as their name suggests, are used to improve digestion and reduce gastrointestinal discomfort (acid reflux, bloating, transit disorders, etc.) . Therapeutic enzymes, used to improve metabolic functions, support immunity, reduce inflammation and pain, etc.

References

F. A. Cotton & G. Wilkinson, *Advanced Inorganic Chemistry*, Wiley, New York, 5^{e}ed. 1990

J. P. Collman, L.S. Hegedus, J. R. Norton & R. G. Finke, *Principles and Applications of Organotransition Metal Chemistry*, Oxford Univ. Press, Oxford, 2^{e}ed. 1987

S. G. Davies, *Organotransition Metal Chemistry: Applications to Organic Synthesis*, Pergamon Press, Oxford, 1982

G. W. Parshall, *Homogeneous Catalysis*, Wiley, 1980

C. Masters, *Homogeneous Transition Metal Catalysis*, Chapman & Hall, London, 1981

G. Wilkinson, F. G. A. Stone & E. W. Abel, *Comprehensive Organometallic Chemistry*, vol. VIII, Pergamon Press, Oxford, 1982

A. Yamamoto, *Organotransition Metal Chemistry. Fundamental Concepts and Applications*, Wiley, New York, 1986.

FSC
www.fsc.org
MIX
Papier | Fördert gute Waldnutzung
FSC® C083411

Kolibri 360 GmbH
Ferdinand-Jühlke-Straße 7
99095 Erfurt, Deutschland
produktsicherheit@kolibri360.de

Druck:
CPI Druckdienstleistungen GmbH
im Auftrag der
Kolibri 360 GmbH
Ferdinand-Jühlke-Str. 7
99095 Erfurt